Geometry, Grade 5

A complete workbook with lessons and problems

By Maria Miller

Contents

Preface

Hello! I am Maria Miller, the author of this math book. I love math, and I also love teaching. I hope that I can help you to love math also!

I was born in Finland, where I also grew up and received all of my education, including a Master's degree in mathematics. After I left Finland, I started tutoring some home-schooled children in mathematics. That was what sparked me to start writing math books in 2002, and I have kept on going ever since.

In my spare time, I enjoy swimming, bicycling, playing the piano, reading, and helping out with Inspire4.com website. You can learn more about me and about my other books at the website MathMammoth.com.

This book, along with all of my books, focuses on the conceptual side of math... also called the "why" of math. It is a part of a series of workbooks that covers all math concepts and topics for grades 1-7. Each book contains both instruction and exercises, so is actually better termed *worktext* (a textbook and workbook combined).

My lower level books (approximately grades 1-5) explain a lot of mental math strategies, which help build number sense — proven in studies to predict a student's further success in algebra.

All of the books employ visual models and exercises based on visual models, which, again, help you comprehend the "why" of math. The "how" of math, or procedures and algorithms, are not forgotten either. In these books, you will find plenty of varying exercises which will help you look at the ideas of math from several different angles.

I hope you will enjoy learning math with me!

Introduction

The problems in this workbook involve lots of drawing. Geometry is a hands-on subject, and many children like that. Moreover, drawing is an excellent means of achieving the conceptual understanding that geometry requires.

Exercises marked with the symbol " " are meant to be done in a notebook or on blank paper.

This workbook starts out with several lessons that review topics studied in previous grades, such as measuring angles, the vocabulary of basic shapes, and how to draw a perpendicular line through a given point on a line. Some fun is included, too, with star polygons.

In the lesson about circles, we learn the terms circle, radius, and diameter. Students draw circles and circle designs using a compass.

Then we go on to classify quadrilaterals and learn the seven different terms used for them. The focus is on understanding the classification, and understanding that attributes defining a certain quadrilateral also belong to all the "children" (subcategories) of that type of quadrilateral. For example, squares are also rhombi, because they have four congruent sides (the defining attribute of a rhombus).

Next, we study and classify different triangles. Students are now able to classify triangles both in terms of their sides and also in terms of their angles. The lesson has several drawing problems and one easy compass-and-ruler construction of an equilateral triangle.

The last focus of this workbook is volume. Students learn that a cube with the side length of 1 unit, called a "unit cube," is said to have "one cubic unit" of volume, and can be used to measure volume. They find the volume of right rectangular prisms by "packing" them with unit cubes and by using formulas. They recognize volume as additive and solve both geometric and real-word problems involving volume of right rectangular prisms.

I hope you find this workbook helpful in teaching math!

Maria Miller, the author

Helpful Resources on the Internet

Use these online resources as you see fit to supplement the main text.

FOR REVIEW OF ANGLES AND POLYGONS

Measuring Angles
Rotate the protractor into position and give your measurement to the nearest whole number.
http://www.mathplayground.com/measuringangles.html

Turtle Pond
Guide a turtle to a pond using commands that include turning him through certain angles and moving him specific distances.
http://illuminations.nctm.org/Activity.aspx?id=3534

Interactive Polygon Crossword Puzzle
Use the clues to help you guess the words that go in the puzzle, and fill it in.
http://www.mathgoodies.com/puzzles/crosswords/ipolygon3.html

Types of Polygons Vocabulary Quiz
In this interactive quiz you have to quickly name different types of polygons based on given clues. For each question you will have only 30 seconds to write your answer!
http://www.math-play.com/types-of-poligons.html

Polygon Matching Game
Many of the polygons included are quadrilaterals.
http://www.mathplayground.com/matching_shapes.html

Free Worksheets for Area and Perimeter
Create worksheets for the area and the perimeter of rectangles/squares with images, word problems, or problems where the student writes an expression for the area using the distributive property. Options also include area and perimeter problems for irregular rectangular areas, and more.
http://www.homeschoolmath.net/worksheets/area_perimeter_rectangles.php

Areas of Rectangular Shapes Quiz
Practice finding the area of rectangular compound shapes with this interactive quiz.
https://www.studyladder.com/games/activity/area-of-irregular-shapes-13136

Circle
This page includes a detailed lesson about circles, as well as interactive exercises to practice the topic.
http://www.mathgoodies.com/lessons/vol2/geometry.html

QUADRILATERALS

Interactive Quadrilaterals
See all the different kinds of quadrilateral "in action." You can drag the corners, see how the angles change, and observe what properties do not change.
http://www.mathsisfun.com/geometry/quadrilaterals-interactive.html

Properties of Quadrilaterals
Investigate the properties of a kite, a rhombus, a rectangle, a square, a trapezoid, and a parallelogram in this dynamic, online activity.
https://www.geogebra.org/m/yekC7cDh

Complete the Quadrilateral
This is a hands-on activity (printable worksheets) where students join the dots to complete quadrilaterals, which helps students learn about the different types of quadrilaterals.
http://fawnnguyen.com/don-stewards-complete-quadrilateral/

Types of Quadrilaterals Quiz
Identify the quadrilaterals that are shown in the pictures in this interactive multiple-choice quiz.
http://www.softschools.com/math/geometry/quadrilaterals/types_of_quadrilaterals/

Quadrilateral Types Practice at Khan Academy
Identify quadrilaterals based on pictures or attributes in this interactive quiz.
https://www.khanacademy.org/math/basic-geo/basic-geo-shapes/basic-geo-classifying-shapes/e/quadrilateral_types

Classify Quadrilaterals Worksheets
Make free printable worksheets for classifying (identifying, naming) quadrilaterals.
http://www.homeschoolmath.net/worksheets/classify_quadrilaterals.php

TRIANGLES

Triangle Shoot
Practice classifying triangles by their angles or by their sides, or identifying types of angles, with this "math splat" game.
http://www.sheppardsoftware.com/mathgames/geometry/shapeshoot/triangles_shoot.htm

Rags to Riches: Classify Triangles by Sides and Angles
Answer multiple-choice questions about classifying triangles by their angles and sides and about angle measures of a triangle in a quest for fame and fortune.
http://www.quia.com/rr/457498.html

Identify Triangles Quiz
A simple multiple-choice quiz about identifying (classifying) triangles either by their sides or angles. You can modify some of the quiz parameters, such as the number of problems in it.
http://www.thatquiz.org/tq-A/?-j1-l34-p0

Interactive Triangles Activity
Play with different kinds of triangles (scalene, isosceles, equilateral, right, acute, obtuse). Drag the vertices and see how the triangle's angles and sides change.
https://www.mathsisfun.com/geometry/triangles-interactive.html

Classify Triangles Worksheets
Make free printable worksheets for classifying triangles by their sides, angles, or both.
http://www.homeschoolmath.net/worksheets/classify_triangles.php

VOLUME

Geometric Solids
Rotate various geometric solids by dragging with the mouse. Count the number of faces, edges, and vertices.
http://illuminations.nctm.org/Activity.aspx?id=3521

Cuboid Exploder and Isometric Shape Exploder
These interactive demonstrations let you see either various cuboids (a.k.a. boxes or rectangular prisms) or various shapes made of unit cubes, and then "explode" them to the unit cubes, illustrating volume.
http://www.teacherled.com/resources/cuboidexplode/cuboidexplodeload.html and
http://www.teacherled.com/resources/isoexplode/isoexplodeload.html

3-D Boxes Activity
Identify how many cubes are in the 3-D shapes in this interactive activity.
http://www.interactivestuff.org/sums4fun/3dboxes.swf

Rectangular Prisms Interactive Activity
Fill a box with cubes, rows of cubes, or layers of cubes. Can you determine a rule for finding the volume of a box if you know its width, depth, and height?
http://illuminations.nctm.org/Activity.aspx?id=4095

Interactivate: Surface Area and Volume
Explore or calculate the surface area and volume of rectangular prisms and triangular prisms. You can change the base, height, and depth interactively.
http://www.shodor.org/interactivate/activities/SurfaceAreaAndVolume/

Decompose Figures To Find Volume - Practice at Khan Academy
Find the volume of irregular 3-D figures by dividing the figures into rectangular prisms and finding the volume of each part.
https://www.khanacademy.org/math/cc-fifth-grade-math/cc-5th-measurement-topic/cc-5th-volume/e/decompose-figures-to-find-volume

Volume Word Problems
Practice solving word problems that involve volume of rectangular prisms.
https://www.khanacademy.org/math/pre-algebra/measurement/volume-introduction-rectangular/e/volume_2

Worksheets for the Volume and Surface Area of Rectangular Prisms
Customizable worksheets for volume or surface area of cubes and rectangular prisms. Includes the option of using fractional edge lengths.
http://www.homeschoolmath.net/worksheets/volume_surface_area.php

FOR FUN

Patch Tool
An online activity where the student designs a pattern using geometric shapes.
http://illuminations.nctm.org/Activity.aspx?id=3577

Interactivate! Tessellate
An online, interactive tool for creating your own tessellations. Choose a shape, then edit its corners or edges. The program automatically changes the shape so that it will tessellate (tile) the plane. Then push the tessellate button to see your creation! Requires Java.
http://www.shodor.org/interactivate/activities/Tessellate

Review: Angles

An angle is a figure formed by two **rays** that have the same beginning point. That point is called the **vertex**. The two rays are called the sides of the angle.

Imagine the two sides as being like two sticks that open up a certain amount. The more they open, the bigger the angle.

An angle can be named (1) after the vertex point, (2) with a letter inside the angle, or (3) using one point on the ray, the vertex point, and one point on the other ray.

We measure angles in degrees. You can use a protractor like the one at the right to measure angles. The angle in blue measures 35 degrees.

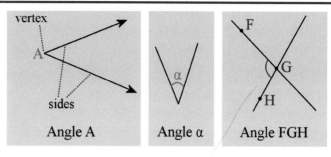

Angle A   Angle α   Angle FGH

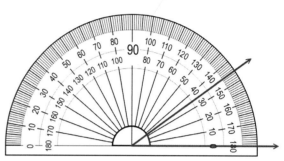

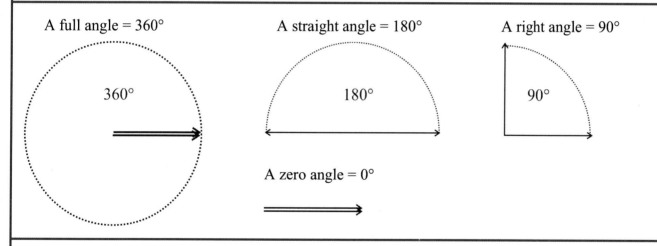

A full angle = 360°   A straight angle = 180°   A right angle = 90°

360°   180°   90°

A zero angle = 0°

Angles that are more than 0° but less than 90° are called **acute** ("sharp") angles.
Angles that are more than 90° but less than 180° are called **obtuse** ("dull") angles.
(Angles that are more than 180° but less than 360° are called *reflex* angles.)

1. Continue the sides of these angles with a ruler.

   Then, measure them with a protractor.

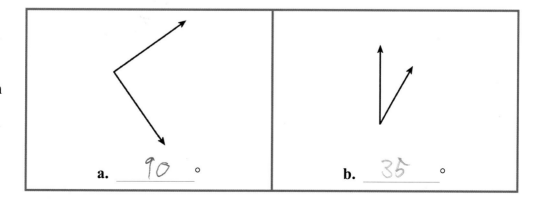

a. _____ 90 °   **b.** _____ 35 °

2. In your notebook, draw:

    a. Any acute angle. Measure it. Label the angle as "An acute angle, xx°."

    b. Any obtuse angle. Measure it. Label the angle as "An obtuse angle, xx°."

3. Draw three dots on a blank paper and join them to form a triangle.
   Draw the dots far enough apart so that the triangle nearly fills the page.
   Then, measure the angles of your triangle.

   The angles of my triangle are: __110__°, __25__°, and __40__°.

   What is the *sum* of these angle measures? __175__°

$$\begin{array}{r} 110 \\ 40 \\ +\ 25 \\ \hline 175 \end{array}$$

4. Draw a horizontal line and mark a point on it. This point will be the vertex of an angle.
   Draw the other side of the angle from the vertex so that the angle measures 76°.

5. Follow the procedure above to draw acute angles with the following measures:
   a. 30° **b.** 60° **c.** 45°

6. Draw obtuse angles with the following measures:
   a. 135° **b.** 100° **c.** 150°

7. Now that you have drawn several angles, *estimate* the angle measure of these angles. Write down the estimates on the top lines. Then measure the angles, and write down the measures on the bottom lines.
   To measure the angles, you will need to continue their sides.

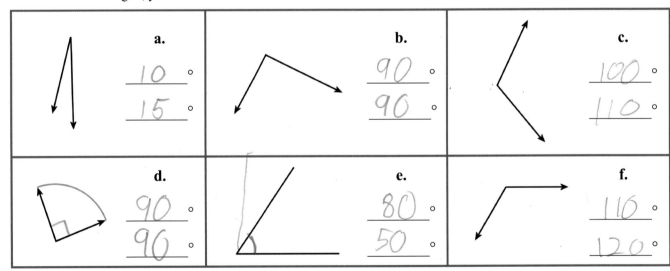

	Important Terms	
• an angle	• an acute angle
• a zero angle	• a right angle
• a straight angle	• an obtuse angle

12

Review: Drawing Polygons

Review these terms for geometric figures:

- A **polygon** – a closed figure made up of line segments.
- A **right triangle** – a triangle with one right angle.
- An **obtuse triangle** – a triangle with one obtuse angle.
- An **acute triangle** – a triangle with all three angles acute.
- A **quadrilateral** – a polygon with four sides.

- A **pentagon** – a polygon with *five* sides.
- A **hexagon** – a polygon with *six* sides.
- A **heptagon** – a polygon with *seven* sides.
- An **octagon** – a polygon with *eight* sides.

- A **vertex** is a "corner" of a polygon.
- A **diagonal** is a line segment drawn from one vertex of a polygon to another (inside the polygon).

These pictures remind you how to use a protractor or a triangular ruler to draw a **perpendicular** line through a given point (a line that is at a right angle with the given line).

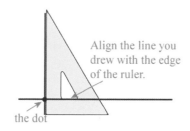

Align the line you drew with the edge of the ruler.

the dot

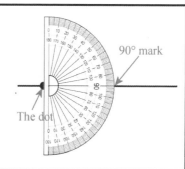

90° mark

The dot

1. Draw any pentagon in your notebook by first drawing five DOTS, and then connecting them with line segments (use a ruler!). Draw the dots kind of randomly around the page and so that your pentagon <u>nearly fills the space</u>. You will need a fairly large pentagon. Don't try to get a regular shape but a pentagon with different side lengths and angles.

   a. Measure all the angles of your pentagon. The angles measure:

   _____°   _____°   _____°   _____°

   , _____ , _____ , _____ , and _____ .

   b. Now draw two diagonals inside your pentagon, dividing it into three triangles. Classify each of those triangles as acute, right, or obtuse.

2. Draw a perpendicular line to the given line through the given point.

   (If you need more practice, repeat this task in your notebook. Start by drawing a line and a point on it.)

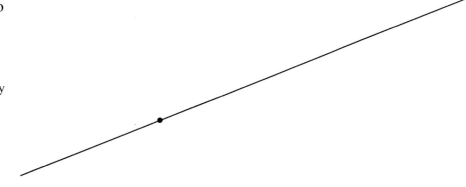

3. Draw a square with 3¼-inch sides. Use a proper tool for drawing perpendicular lines.

   (1) Draw a long line, longer than necessary. Mark on it the first 3¼-inch side.

   (2) Now draw two perpendicular lines from the two endpoints of the 3¼-inch side.

   (3) Measure the other two 3¼-inch sides. Draw dots where the two vertices will be.

   (4) Draw in the last side of your square.

How to draw a triangle with two given angle measurements.

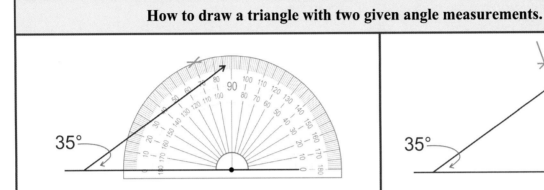

Let's say you have already drawn a 35° angle, and the second angle is supposed to be 70°. The image shows you how to place your protractor so you can measure and mark the 70° angle.

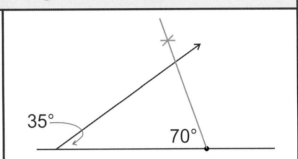

Then remove the protractor and draw the third side of the triangle.

---

4. **a.** Draw a triangle with 50° and 75° angles.

   b. Measure the third angle. It measures _____°.

   c. What is the sum of the three angle measures?

5. **a.** Draw a triangle with 110° and 35° angles.

   b. Measure the third angle. It measures _____°.

   c. What is the sum of the three angle measures?

Important Terms		
• a right triangle	• vertex	• pentagon
• an acute triangle	• diagonal	• hexagon
• an obtuse triangle	• perpendicular	• heptagon
• polygon	• quadrilateral	• octagon

Star Polygons (*optional*)

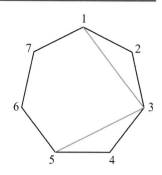

1. This picture shows a regular heptagon where every *other* vertex is connected with a line segment (skipping one vertex in between).

   Continue drawing diagonals in such a manner. The shape you will get is called a *star polygon*, and specifically a *heptagram*.

2. Make star polygons.

a. This is a regular nonagon. Make a nonagram connecting every four vertices (skipping 3).

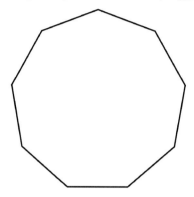

b. This is a regular pentagon. Make a pentagram connecting every other vertex (skipping one).

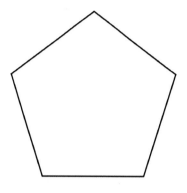

c. This is a regular octagon. Make an octagram connecting every three vertices (skipping 2).

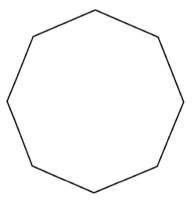

d. This is a regular heptagon. Make a heptagram connecting every three vertices (skipping 2).

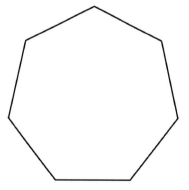

e. This is a regular decagon. Make a decagram connecting every three vertices (skipping 2).

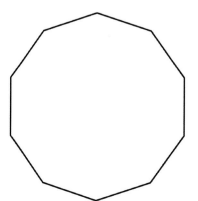

f. Make a nonagram connecting every second vertex (skipping 1).

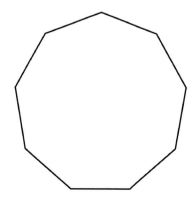

3. You can make your own star polygons here. Experiment!

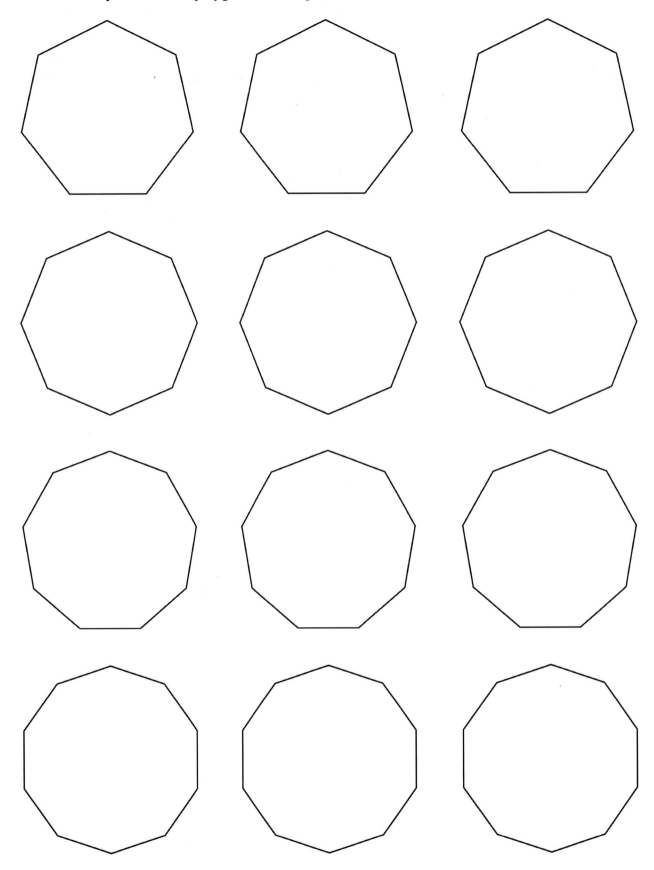

Circles

These figures are round, but they are not circles.

These are ovals. They are symmetric and round, but they are still not circles. Why not?

What makes a circle?
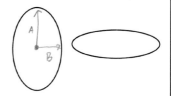

The difference between other round figures and circles is this:

In a circle, the <u>distance</u> from the **center point** to the actual circle line, or **circumference of the circle**, remains the same.

This distance is called the **radius** of the circle.

In other words, all the points on the circumference are **at the same distance from the center point**.

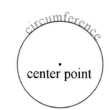

The distance from the center point to any point on the circumference is called the **radius**.

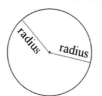

A line through the center point is called a **diameter**.

1. Draw a radius or a diameter from the given point. Use a ruler. Look at the example.

Here, a radius is drawn from the given point.

a. Draw a radius from the given point.

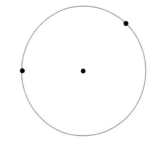

b. Draw a radius from each of the given points.

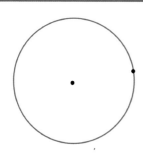

c. Draw a diameter from the given point.

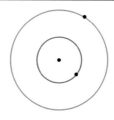

d. Draw a diameter for the smaller circle and a diameter for the bigger circle from the given points.

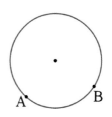

e. Draw a radius from the point A and a diameter from the point B.

2. Learn to use a compass to draw circles.

   a. Draw many circles with the compass.

   b. Now, set the radius on the compass to be 3 cm, and draw a circle. You can do that by placing the compass next to a ruler, and adjusting the radius of the compass until it is 3 cm as measured by the ruler. Some compasses show the radius for you, so you won't need a ruler.

   c. Draw a circle with a radius of 5 cm.

   d. Draw a circle with a radius of 1 ½ in.

3. **a.** Draw two diagonals into this square. Draw a point where they cross (the center point of the square). Now, erase the lines you drew, leaving the point.

   b. Draw a circle *around* the square so that it touches the vertices of the square. Use the point you drew in (a) as the center point.

   c. Fill in: The _____ of the circle has the same length as the diagonal of the square.

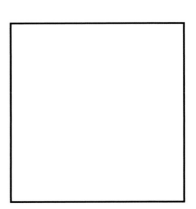

4. **a.** Draw a circle *inside* this square so that it touches the sides of the square but will not cross over them.

   b. Fill in: The _____ of the square has the same length as the diameter of the circle.

   You can repeat or practice exercises #3 and #4 in your notebook.

5. **a.** Draw a circle with center point (5, 6) and a radius of 2 units. Use a compass.

   b. Draw another circle with the same center point, but double the radius.

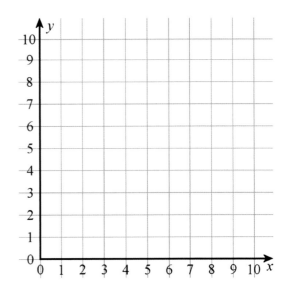

6. Draw these figures using a compass and a ruler in your notebook. Your figures don't have to be the same size as these; they just need to show the same pattern. *See hints at the bottom of this page. Optionally, you can also draw these in drawing software.*

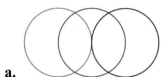
a.

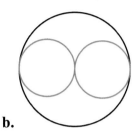
b.

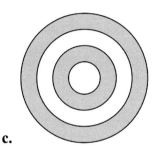
c.

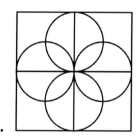
d.

See **http://homeschoolmath.blogspot.com/2013/02/geometric-art-project-seven-circle.html** for one more circle design and art project!

a. Hint: Draw a line. Then, draw the three center points on it, equally spaced.

b. Hint: First, draw the three center points for the three circles, equally spaced. What is the radius of the big circle compared to the radius of the small ones?

c. Hint: What pattern is there in the radii of these circles? These circles are called concentric circles because they share the same center point.

d. Hint: You need to draw the outer square first. Then measure and divide it into quarters. Measure to draw the center points of the circles (they are midpoints of the sides of the smaller squares).

New terms to remember:	
• circle	• radius
• circumference	• diameter

19

Quadrilaterals

Quadrilaterals are polygons with four sides (*quadri-* = four, *lateral* = referring to a side). You already know about the three quadrilaterals below.

Note: If two sides of a quadrilateral have the same length, they are said to be **congruent**.

1. A **parallelogram** has two pairs of parallel sides.	2. A **rectangle** has four right angles.	3. A **square** is a rectangle with four congruent sides.

We can organize these three quadrilaterals in a **tree diagram** (on the right).

Start "reading" the tree diagram from the top, beginning with the parallelogram. The next figure, the rectangle, is like a "child" to the parallelogram. If the parallelogram is the parent of the family, then its child has the same "family name" because it belongs to the parallelogram family.

Why? Because a rectangle also has two pairs of parallel sides. So it, too, is a parallelogram! Additionally, its angles are right angles, so a rectangle has something *more* than a parallelogram does.

Similarly, a square is like a "child" of the rectangle. The square has the same properties as its "parent" and "grandparent": the square is also a rectangle, and it is also a parallelogram. Additionally, all of its sides are congruent.

A Tree Diagram

4. A **rhombus** is a parallelogram that has four congruent sides (a diamond).

Now let's add a new "family member" to the diagram: the *rhombus*.

The rhombus is a parallelogram, so it belongs under the parallelogram in the tree diagram.

It shares something with the square, as well. Both have four congruent sides. This means the square goes *under* the rhombus. But the rhombus and the rectangle do *not* share characteristics (other than both being quadrilaterals).

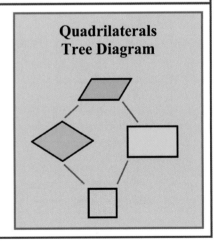

Quadrilaterals Tree Diagram

1. Draw a quadrilateral that has **four right angles** and **one side 2 inches long**. Can you draw only *one* kind of quadrilateral like that, or can you draw *several* kinds that all look different? (Compare your results with those of your classmates.)

2. A *regular* quadrilateral would be a quadrilateral whose sides are all congruent and whose angles all have the same measure. What is the usual name for such a figure?

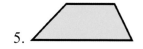 5.	 6.	7.
A **trapezoid** has *at least* one pair of parallel sides. It may have two!	A **kite** has two pairs of congruent sides that touch each other. *The single tick marks show the one pair of congruent sides, and the double tick marks show the other pair.*	In a **scalene** quadrilateral, all sides are of different lengths (no two sides are congruent).

3. Each quadrilateral below is either a parallelogram, a rhombus, a trapezoid, or a kite. Write their names.

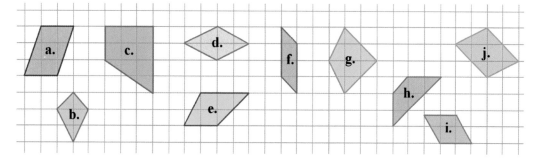

a. _____   b. _____

c. _____   d. _____

e. _____   f. _____

g. _____   h. _____

i. _____   j. _____

4. In the grid below:

   a. Join the dots to make a parallelogram.   **b.** Draw two more different parallelograms.

   c. Draw a rhombus.   **d.** Draw a kite.   **e.** Draw a scalene quadrilateral.

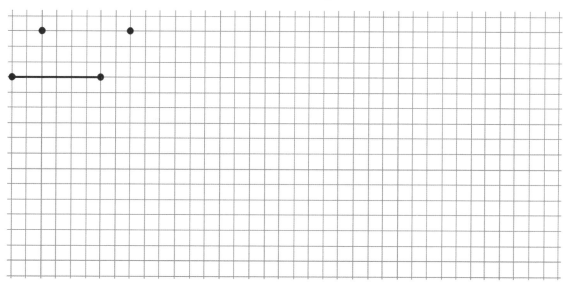

5. Our tree diagram is now complete with seven different kinds of quadrilaterals! Name each type of quadrilateral in the diagram.

6. Answer the questions. The tree diagram will help.

   a. Is a rhombus also a kite?

   b. Is a square also a kite?

   c. Is a rectangle also a kite?

   d. Is a square also a trapezoid?

   e. Is a parallelogram also a kite?

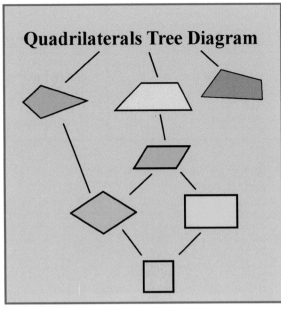

Quadrilaterals Tree Diagram

7. Draw a trapezoid so that its parallel sides measure 3 in and 1 ¾ in. Can you draw several such trapezoids that are not identical?

8. The sides of a parallelogram measure 2 in, 3 in, 2 in, and 3 in. Is it also a kite? A rhombus? A trapezoid?

9. A trapezoid's sides measure 5 cm, 3 cm, 8 cm, and 3 cm. Is it also a kite? A parallelogram? A rhombus?

10. Janine is supposed to draw a quadrilateral with each side 1½ inches long. Can she draw only one kind, or can she draw several different-looking ones? Explain.

11. A certain quadrilateral has two pairs of congruent sides and also two pairs of parallel sides. What kind of quadrilateral is it?

12. Solve the quadrilaterals puzzle and uncover a message!

Quadrilateral	Letter
I have one right angle. I look like a toy that comes back to you.	
I have the biggest area around here!	
I am the little rhombus!	
Two (and only two) of my sides are parallel. The other two are congruent.	
There's nothing congruent or parallel about me.	
I'm the bigger parallelogram with four congruent sides.	
I am the big rhombus!	
I've got right angles and two pairs of congruent sides.	
I have four right angles and four congruent sides!	
Two—and only two—of my sides are parallel. None are congruent.	
They call me a rectangle.	
I'm one of the parallelograms in the bunch with a vowel inside me.	
I have two pairs of parallel sides, but not all of my sides are congruent.	
You can't find a more regular quadrilateral than me.	
While I do have two parallel sides, you cannot draw a line of symmetry through me.	
Trapezoid is my name, but I don't have any congruent sides.	
Scalene—yep, that's me!	
None of my sides are parallel to each other, but I have two pairs of congruent sides.	
I'm the diamond-shape again... the bigger one.	
If you stretch me, I'll be a rectangle.	
Oops! I don't even belong to this group!	
	!

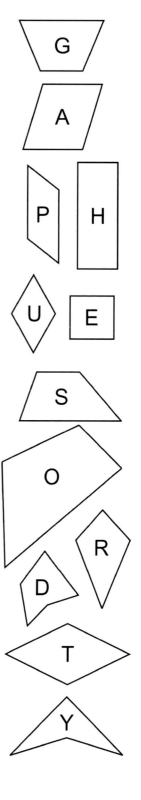

Equilateral, Isosceles, and Scalene Triangles

Classification according to sides	
If all three sides of a triangle are congruent (the same length), it is called an **equilateral triangle**. "*Equi-*" refers to things that are the same or equal, and "*lateral*" refers to sides. Think of it as a "same-sided" triangle. 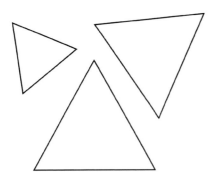	If only *two* of a triangle's sides are congruent, then it is called an **isosceles triangle**. Think of it as a "same-legged" triangle, the "legs" being the two sides that are the same length. MARK the two congruent sides of each isosceles triangle: 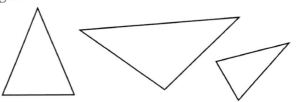
	Lastly, if none of the sides of a triangle are congruent (all are different lengths), it is a **scalene triangle**. 

1. Classify the triangles by the lengths of their sides as either equilateral, isosceles, or scalene.

   You can mark each triangle with an "*e*," "*i*," or "*s*" correspondingly.

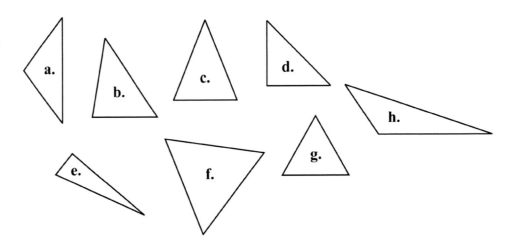

2. Plot the points (0, 0), (3, 5), (0, 5) , and connect them with line segments to form a triangle.

   Classify your triangle by its sides.
   Is it equilateral, isosceles, or scalene?

   _____

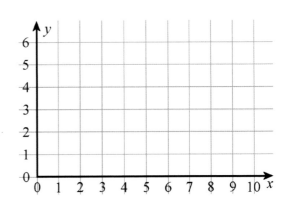

24

3. Classify the triangles as "acute," "right," or "obtuse" (by their angles), and also as "equilateral," "isosceles," or "scalene" (by their sides).

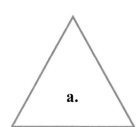 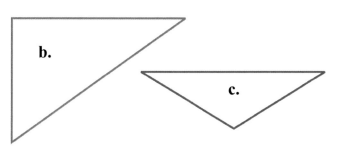 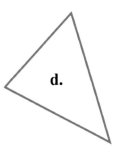

Triangle	Classification by the angles	Classification by the sides
a.		
b.		
c.		
d.		

4. Plot the points, and connect them with line segments to form two triangles. Classify the triangles by their angles <u>and</u> sides.

Triangle 1: (0, 0), (4, 0), (0, 4)

_____ and

_____

Triangle 2: (5, 5), (1, 8), (9, 4)

_____ and

_____

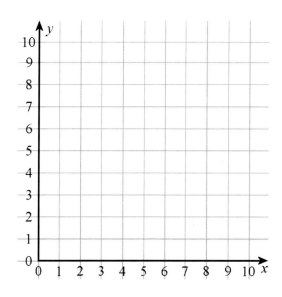

25

5. Fill in the missing parts in this tree diagram classification for triangles.

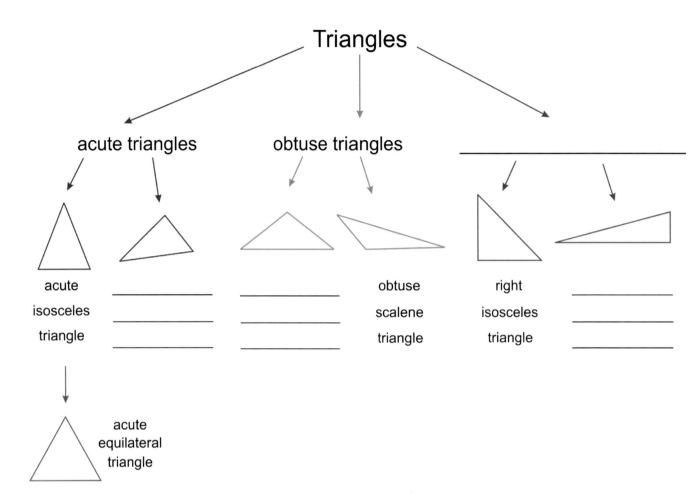

Triangles

acute triangles   obtuse triangles   _____

acute
isosceles   _____   _____   obtuse   right
triangle   _____   _____   scalene   isosceles   _____
_____   _____   triangle   triangle   _____
_____

acute
equilateral
triangle

6. Sketch an example of the following shapes. You don't need to use a ruler.

  a. an obtuse isosceles triangle

  b. an obtuse scalene triangle

  c. a right scalene triangle

7. Plot in the grid

  a. a right isosceles triangle

  b. an acute isosceles triangle.

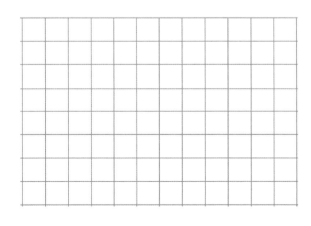

8. Make a guess about the angle measures in an equilateral triangle: _____°
   Measure to check.

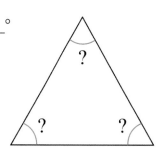

9. **a.** Could an equilateral triangle be a right triangle?
   If yes, sketch an example. If not, explain why not.

   b. Could a scalene triangle be obtuse?
   If yes, sketch an example. If not, explain why not.

   c. Could an acute triangle be scalene?
   If yes, sketch an example. If not, explain why not.

10. State whether or not it is possible to draw the following figures. (You don't have to draw any.)

   a. an obtuse equilateral triangle

   b. a right equilateral triangle

   c. an acute isosceles triangle

11. Measure all the angles in these isosceles triangles. Continue their sides, if necessary.
    Mark the angle measures near each angle.

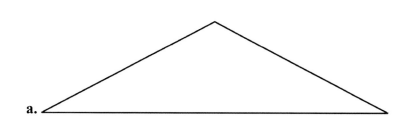

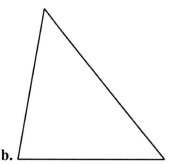

12. **a.** Draw any isosceles triangle.
    Hint: Draw any angle. Then, measure off the two congruent sides, making sure they have the same length. Then draw the last side.

   b. Measure the angles of your triangle. They measure _____°, _____°, and _____°.

13. Based on the last two exercises, can you notice something special about the angle measures?

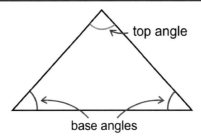

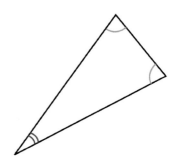

There are two angles in an isosceles triangle that have the SAME angle measure. They are called the **base angles**.

The remaining angle is called the **top angle**.

Can you find the top angle and the base angles in this isosceles triangle?

14. Draw an isosceles triangle with 75° base angles. (The length of the sides can be anything.)
Hint: Start by drawing the base side (of any length). Then, draw the 75° angles.

15. **a.** Draw an isosceles right triangle whose two sides measure 5 cm.
Hint: Draw a right angle first. Then, measure off the 5-cm sides. Then draw in the last side.

b. How long is the third side?

c. What is the measure of the base angles?

16. Draw a scalene obtuse triangle where one side is 3 cm and another is 7 cm.
Hint: Draw the 7-cm side first, then the 3-cm side forming any obtuse angle with the first side.

a. Draw two isosceles triangles with a 50° top angle. Your two triangles should not be identical.

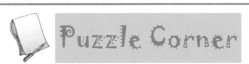

b. What is the angle measure of the base angles?

New Terms
• *equilateral triangle*   • *isosceles triangle*   • *scalene triangle*

Area and Perimeter Problems

Example 1. Find the area of the shaded figure.

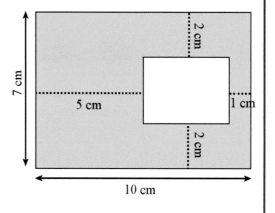

The easiest way to do this is:
(1) Find the area of the larger outer rectangle.
(2) Find the area of the white inner rectangle.
(3) Subtract the two.

1. The area of the large rectangle is 7 cm × 10 cm = 70 cm^2.

2. We find the *sides* of the white rectangle by subtracting.

   The longer side of the white rectangle is
   10 cm − 5 cm − 1 cm = 4 cm.
   The shorter side is 7 cm − 2 cm − 2 cm = 3 cm.

   So, the area of the white rectangle is 4 cm × 3 cm = 12 cm^2.

3. Now we subtract to find the shaded area: 70 cm^2 − 12 cm^2 = 58 cm^2.

1. **a.** Find the area of the white rectangle.
   All lines meet at right angles.

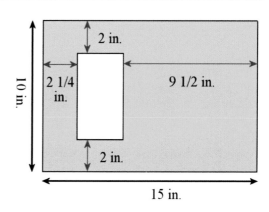

   b. Find the area of the shaded figure.

2. The image on the right shows a picture frame.
   Find the area of the actual frame (that is, of the shaded part).
   All lines meet at right angles.

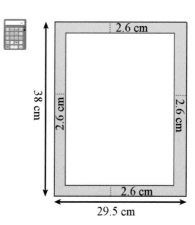

29

Example 2. Find the perimeter of the figure.

We need to find the length of *each* side and then add the lengths. Start, for example, at the side marked with 1, then go to the side marked with 2, then to side 3, and so on, until you have "traveled" all the way around the figure.

Side 1 is 3 cm. Side 2 is 2 cm. Side 3 is 5 cm.
The total perimeter is:

3 cm + 2 cm + 5 cm + 5 cm + 4 cm + 1 cm + 4 cm + 4 cm = 28 cm.

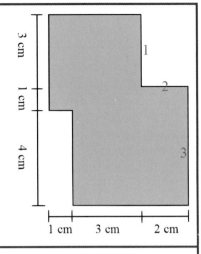

Example 3. Find the area of the figure.

Divide the figure into rectangles by drawing in it some additional lines.

Rectangle 1 has an area of 4 cm × 4 cm = 16 cm^2.

Rectangle 2 has an area of 3 cm × 4 cm = 12 cm^2.

Rectangle 3 has an area of 2 cm × 5 cm = 10 cm^2.

The total area is: 16 cm^2 + 12 cm^2 + 10 cm^2 = 38 cm^2.

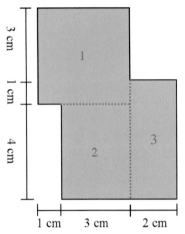

3. Find the area and the perimeter of this figure.
   All lines meet at right angles.

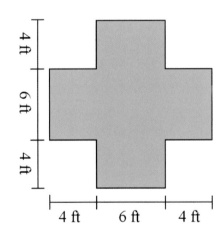

4. The perimeter of a rectangle is 42 cm.
   If the long side of the rectangle is 11 cm,
   how long is the shorter side?

5. Find the area and the perimeter of this figure.
All lines meet at right angles.
The dimensions are given in centimeters.

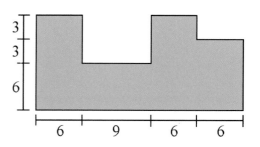

6. One side of a rectangular field measures 330 ft.
A farmer fenced it with 910 ft of fencing.
How long is the other side of the field?

7. The perimeter of a square is ½ mile.

a. How long is one side of the square, in miles?
Draw a sketch to help you.

b. How long is one side of the square, in *feet*?

8. Find the area and the perimeter of this figure.
All lines meet at right angles.
The dimensions are given in inches.

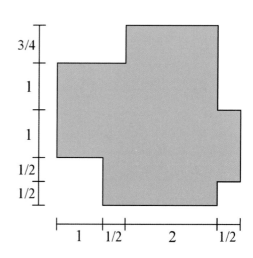

Volume

The **volume** of an object has to do with how much SPACE it takes up or occupies.

You have measured the volume of liquids using measuring cups that use ounces or milliliters. If we need to know the volume of a big object, such as a room, we cannot pour water into it to measure it with measuring cups. Instead, we use cube-shaped units or **cubic units**, and we simply check or calculate how many cubic units fit into the object.

 This little cube is **1 cubic unit.**

The volume of the figure on the right is six cubic units: V = 6 cubic units. Notice that one cube is not visible.

1. Find the volume of these figures in cubic units. "V" means volume.

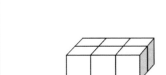

a. V = _____ cubic units

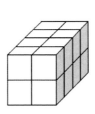

b. V = _____ cubic units

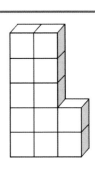

c. V = _____ cubic units

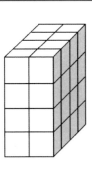

d. V = _____ cubic units

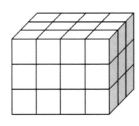

e. V = _____ cubic units

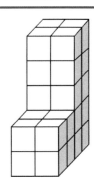

f. V = _____ cubic units

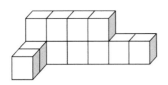

g. V = _____ cubic units

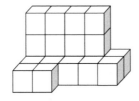

h. V = _____ cubic units

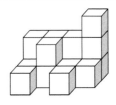

i. V = _____ cubic units

If each edge of this cube measures 1 cm, then the volume of the cube is **1 cubic centimeter**. This is abbreviated as **1 cm³**.

If each edge of the cube is 1 foot, its volume is **1 cubic foot.**

V = 1 cu. ft. or 1 ft³

If each edge of the cube is 1 inch, its volume is **1 cubic inch**.

V = 1 cu. in. = 1 in³

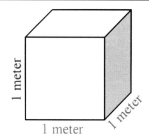

If each edge of the cube is 1 meter, its volume is **1 cubic meter.**

V = 1 m³

In general, if the edges of the cube are in certain units (such as inches, feet, centimeters, or meters), then the volume will be in corresponding *cubic* units.

If no unit is given for the edge lengths, we use the word "unit" for the lengths of the edges, and "cubic unit" for the volume. This "box" has a volume of 18 cubic units.

2. Find the total volume of each figure when the edge length of the little cube is given. Remember to include the unit!

The edge of each cube is 1 in.

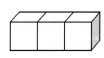

a. V = _____ *3 in³* _____

The edge of each cube is 1 ft.

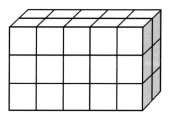

b. V = _____

The edge of each cube is 1 cm.

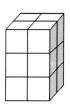

c. V = _____

The edge of each cube is 1 m.

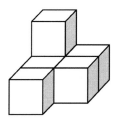

d. V = _____

The edge of each cube is 1 cm.

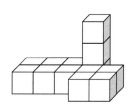

e. V = _____

The edge of each cube is 1 in.

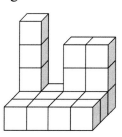

f. V = _____

This figure is called a **rectangular prism.** It is also called *a cuboid.* It is simply a box with sides that meet at right angles.

Many people call the **three dimensions** that we measure "length," "width," and "height." Here we will use "width," "depth," and "height."

The **width** will be the dimension that runs left to right.
The **depth** will be the dimension that points away from you—into the paper, so to speak.
The **height** will be the dimension pointing "up" in the figure.

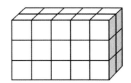

A way to find the volume of a rectangular prism by calculating

1) Can you figure out a way to find the number of cubes in the *bottom* layer of this rectangular prism *without* counting?

   You can multiply 5 × 2 = 10, which means multiplying the *width* and the *depth*. The bottom layer has 10 cubic units.

2) After that, there is a way to easily find the *total* number of cubes in the rectangular prism (its volume). Can you figure that out?

   Take the number of cubes in the bottom layer, and **multiply that by how many layers there are** (the *height*). There are 10 cubes in the bottom layer, and 3 layers. We get 10 × 3 = 30 cubic units.

---

3. Find the volume of these rectangular prisms by finding the amount of cubic units in the bottom layer and multiplying that by the height (how many layers there are).

	a.	**b.**	**c.**	**d.**
Cubes in the bottom layer	*8*			
Height	*4*			
Volume	*32*			

4. If each little cube is 1 cubic inch, what is the total volume of the outer box?

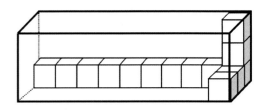

Notice what we did in these two steps:

(1) We multiplied the <u>width</u> and the <u>depth</u> to find the number of cubes in the bottom layer. **Multiplying the width and the depth** also gives us **the area of the bottom** (A_b)! For example, the bottom area of this cuboid is $4 \times 3 = 12$ square units.

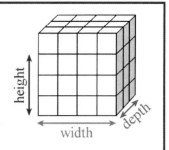

(2) We multiplied what we got from step 1 by <u>height</u>.

We ended up multiplying the bottom area by the height.
Or, looking at it in another way, we multiplied the width, the depth, and the height.

From that we get two **formulas** for the volume of a rectangular prism:

1. $V = w \times d \times h$ (volume is width × depth × height)

2. $V = A_b \times h$ (volume is area of the bottom × height)

5. Write the width, height, and depth of these rectangular prisms. Lastly, multiply those three dimensions to find the volume.

	a.	b.	c.	d.
Width:				
Depth:				
Height:				
Volume:				

6. Find the volume of the rectangular prisms above *if* their top layer was removed. Use cubic units. Use the formula $V = w \times d \times h$.

a. V = _____ × _____ × _____ = _____ cubic units

b. V = _____ × _____ × _____ = _____ cubic units

c. V = _____ × _____ × _____ = _____ cubic units

d. V = _____ × _____ × _____ = _____ cubic units

Now we can explain where the little raised "3" (the exponent) in cubic units comes from.

To find the volume of this cube, we multiply its width, the height, and the depth. This means multiplying 1 cm × 1 cm × 1 cm. Not only do we multiply number 1 by itself three times—we also multiply the *unit centimeter* (cm) by itself <u>three</u> times. The little "3" in cm^3 shows that.

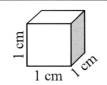

7. **a.** Sketch a rectangular prism with a volume of 4 × 2 × 6 cubic units.

b. Sketch a rectangular prism with a volume of 3 × 3 × 3 cubic units.

c. Sketch a rectangular prism with a volume of 2 × 5 × 4 cubic units.

8. Chris and Mia drew these rectangular prisms to match the expression 5 × 3 × 2. Who is right?

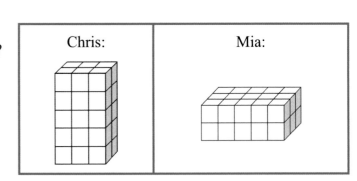

9. To calculate the volume of this kind of figure, think of it as consisting of *two* rectangular prisms. We calculate the volume of each separately, and then add. Which expression below matches the volume of this figure?

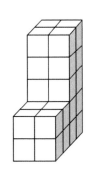

a. 2 × 3 × 2 + 2 × 2 × 3

b. 2 × 2 × 2 + 2 × 2 × 3

c. 2 × 2 × 2 + 2 × 2 × 5

Volume of Rectangular Prisms (Cuboids)

Study the two formulas for the volume of a rectangular prism:

1. $V = w \times d \times h$ (volume is width × depth × height)
Some people use width, <u>length</u>, and height instead.

2. $V = A_b \times h$ (volume is area of the bottom × height)

The width, depth, and height need to be in the <u>same</u> kind of unit of length (such as meters). The volume will then be in corresponding cubic units (such as cubic meters).

Example 1. A room measures 12 ft by 8 ft, and it is 8 ft high. What is the volume of the room? What is the area of the room?

To find the area, we simply multiply the two given dimensions: A = 12 ft × 8 ft = 96 ft².
To find the volume, we can multiply the area by the height: V = 96 ft² × 8 ft = 768 ft³.

1. **a.** Find the volume of a box that is 2 inches high, 5 inches wide, and 7 inches deep. Include the units!

V = ___5 in___ × _____ × _____ = _____

b. Find the area and volume of a room that is 25 ft × 20 ft, and 9 feet high. Include the units!

A = _____ × _____ = _____

V = _____ × _____ × _____ = _____

2. Find the volume of a box that is

a. 20 cm wide, 30 cm deep, and half a meter high.
Note: you will need to convert the last measurement into centimeters before calculating the volume.

b. 16 square inches on the bottom, and 6 inches tall.

3. The volume of this box is 30 cm³. What is its depth?

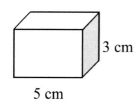

3 cm

5 cm

4. *Optional.* Measure the width, height, and depth of a dresser and/or a fridge. Find out its volume.

Volume is **additive**. What we mean by that is that we can ADD to find the total volume of a shape that is in several parts.

To find the total volume of the shape on the right, first find the volume of the top box, then the volume of the bottom box, and add the two volumes.

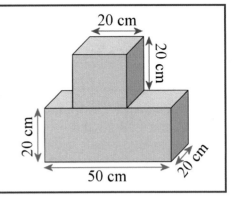

5. Find the total volume of the shape in the teaching box above.

6. This is a two-part kitchen cabinet. Its height is 2 ft and depth 1 ft. One part is 5 ft long, and the other is 4 ft long.

   a. Mark the given dimensions in the picture.

   b. Calculate the volume.

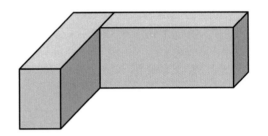

7. Design a box (give its width, height, and depth) with a volume of

   a. 64 cubic inches

   b. 1,200 cubic centimeters

8. The length and width of a rectangular box are 5 inches and 6 inches. Its volume is 180 cubic inches. How tall is it?

9. Find the volume of this building.

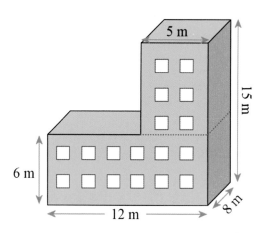

10. Find the total volume.

11. The picture shows an aquarium that is 1/4 filled with water.

    a. Find the total volume of the aquarium.

 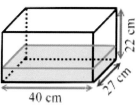

    b. Find the volume of the water in it.

12. John's room is 12 ft × 18 ft, and it is 9 ft high. The family plans to *lower* the ceiling by 1 foot.

    a. What will the volume of the room be after that?

    b. How much volume will the room lose?

13. **a.** Find the volume of this two-part bird cage.

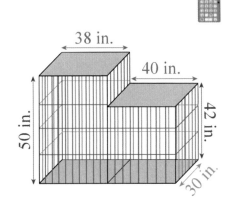

    b. One cubic foot is 1,728 cubic inches.
      Convert your answer from (a) into cubic feet, to three decimals.

Puzzle Corner

The volume of the larger cube is 1,000 cubic inches. The edge length of the smaller cube is half of the edge length of the larger cube.

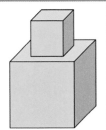

What is the combined volume of the two cubes?

A Little Bit of Problem Solving

1. Fill in the table, continuing the patterns.

   a. What is the area of the 11th square?

   b. What is the number of the square with an area of 10,000 m²?

Square number	Length of Side	Area
1	2 m	4 m²
2	4 m	
3	6 m	
4		
5		
6		

2. A wall is 16 feet wide and 10 feet high. It has one 3.5 ft × 4.5 ft window in it.

   a. Draw a sketch.

   b. Find the area of the window.

   c. What is the area of the actual wall (not including the window)?

   d. A gallon of paint covers 350 square feet of wall.
   How many whole gallons do you need to paint the wall?

   e. If instead of gallons, you buy paint by quarts, how many whole
   quarts of paint do you need to paint the wall?

3. This aquarium's dimensions are 4 ft × 3 ft × 3 ft.

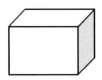

   a. How many cubic feet of water would fit in it?

   b. You fill it 8/9 full. How many cubic feet of water is that?

4. The volume of this cube is 8 cubic inches.
   How long is its edge?

5. What part of the whole design does the highlighted area make up?
   (Remember to simplify your fraction.)

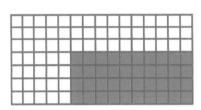

6. A certain room measures 15 ft × 20 ft. What part
   of the floor would a 5 × 4 ft carpet cover?

7. **a.** Draw a right triangle with 5 cm and 12 cm sides, using these steps:

   1. Draw a long line.
   2. Measure the 12-cm side on the line, marking it with two dots.
   3. Draw a perpendicular line through one of the dots. Use a proper tool!
   4. On the second line, measure and mark the 5-cm side.
   5. Draw in the third side of the triangle now.

   b. Find the perimeter of your triangle, in centimeters.

Review

1. Measure all the angles of the triangles. Then classify the triangles.

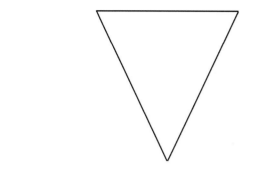

  a. Angles: _____ °, _____ °, _____ °

    Acute, obtuse, or right?

    Equilateral, isosceles, or scalene?

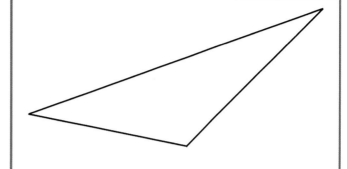

  b. Angles: _____ °, _____ °, _____ °

    Acute, obtuse, or right?

    Equilateral, isosceles, or scalene?

2. **a.** Draw an isosceles triangle with 50° base angles and a 7 cm base side (the side between the base angles).

  b. Measure the top angle.

    It is _____ ° .

  c. Find the perimeter of your triangle in millimeters.

3. Find the perimeter and area of this figure. All measurements are in inches.

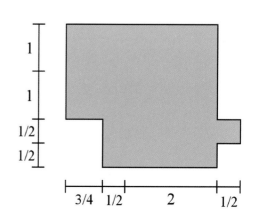

43

4. Name the different types of quadrilaterals.

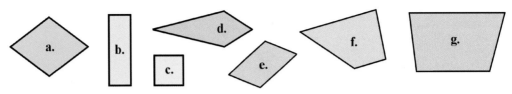

a.   **b.**

c.   **d.**

e.   **f.**

g.

5. Name the quadrilateral that...

   a. is a parallelogram and has four right angles.

   b. is a parallelogram and has four sides of the same length.

   c. has two parallel sides and two sides that are not parallel.

6. **a.** What is this shape called?

   b. Draw enough diagonals inside the shape to divide it into triangles.

   c. Number each of the triangles.

   d. Classify each triangle according to its sides (equilateral, isosceles, scalene) and according to its angles (acute, obtuse, right).

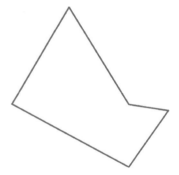

7. **a.** Draw an isosceles obtuse triangle.

   b. Draw a scalene acute triangle.

8. **a.** Draw a circle with its center at (2, 3) and a radius of 2 units. Use a compass.

   b. Draw another circle with its center at (6, 5) and a *diameter* of 8 units.

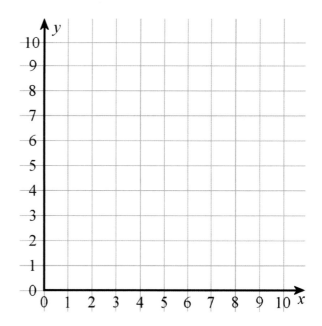

9. Find the volume of this rectangular prism, if…

   a. …the edge of each little cube is 1 inch.

   b. …the edge of each little cube is 2 inches.

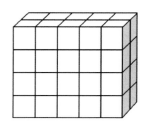

10. What is the height of this box, if its bottom dimensions are 2 cm × 4 cm and its volume is 32 cubic centimeters?

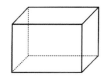

11. A gift box is 6 inches wide, 3 inches deep, and 2 inches tall. How many of these boxes do you need to have a total volume of 108 cubic inches?

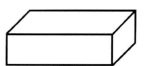

Puzzle Corner   The area of the bottom face of a cube is 16 cm². What is its volume?

45

Geometry, Grade 5
Answer Key

Review: Angles, p. 11

1. a. 90°   b. 27°

2. Check the students' answers. The acute angle should measure less than 90°, and the obtuse angle should measure more than 90°.

3. Check the students' triangles. The sum of the angle measures should be about 180°. Due to measuring errors it may also end up being 179° or 181°.

4.
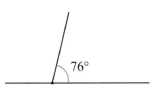
76°

5. a.
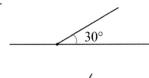
30°

   b.
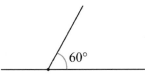
60°

5. c.
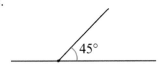
45°

6. a.
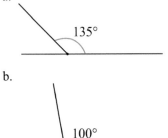
135°

   b.
100°

   c.
150°

7. The estimates will vary. The exact measures are:
   a. 15°   b. 94°   c. 114°   d. 86°   e. 55°   f. 121°

Review: Drawing Polygons, p. 13

1. a. Answers will vary.
   b. Answers will vary.

2.
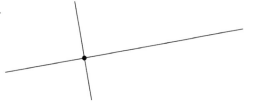

3. (1) Draw a long line. Mark on it the first 3 ¼-inch side.

3 1/4 in.

(2) Now draw two perpendicular lines from the two endpoints of the 3¼-inch side.

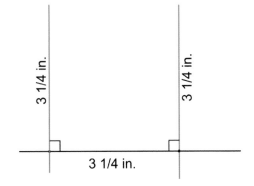

3 1/4 in.

3 1/4 in.

3 1/4 in.

Review: Drawing Polygons, cont.

(3) Measure the other two 3¼-inch sides. Draw dots where the two vertices will be.

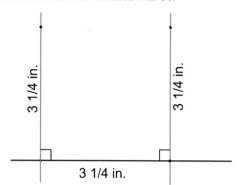

(4) Draw in the last side of your square.

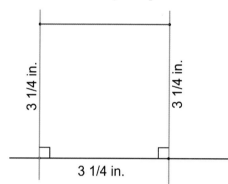

4.

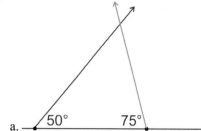

a.

b. The third angle measures 55°. What the student measures may be slightly different from that, depending on how accurate the drawing is.

c. The angle sum should be 180° or close to it.

5.

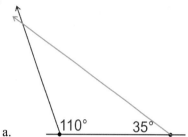

a.

b. The third angle measures 35°. What the student measures may be slightly different from that, depending on how accurate the drawing is.

c. The angle sum should be 180° or close to it.

Star Polygons, p. 15

1.

2.

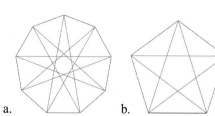

a.   b.

2.

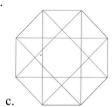 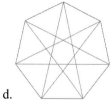

c.   d.

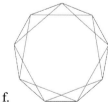

e.   f.

1.

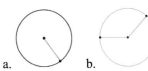

a.   b.

c.   d.   e.

2. a. The student should practice drawing circles
      until he/she can use the compass.
   b. c. d. Check the student's drawings.

3. a.- b. See the image below.

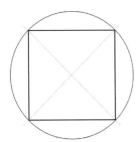

   c. The _diameter_ of the circle

4. a. See the image below. Again, draw the two diagonals
      first, and use the point where they intersect as the
      center point of the circle.

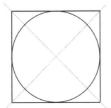

   b. The _side_ of the square.

5.

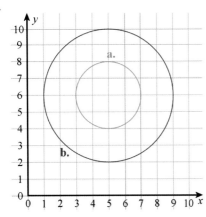

6. a. Draw a straight line, and on it, three points, equally
      spaced.

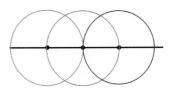

   For example, you can choose the points to be 3 cm
   apart. That 3 cm will be the radius, and the points
   will be the center points for the circles.

   After you have the points, draw the 3 circles. Use the
   distance between 2 points as the radius.

6. b. Draw a line and five dots, evenly spaced
      (for example, 3 cm apart).

   The second and fourth of these dots are
   the center points of the two small circles.

   Draw the two little circles using the chosen radius.

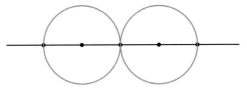

   Lastly, draw the large circle using the middle dot
   as center point and the distance to the outer points
   as a radius. The radius of the small circles is exactly
   half of that of the big circle.

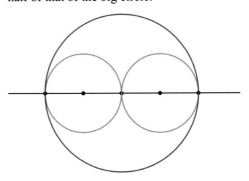

Circles, cont.

6. c. These circles are concentric: they have the same center point. The radius increases by the same amount each time. Simply draw a dot for the center point, and choose the first radius. For example, if the first circle has radius 1 cm, the next one has radius 2 cm, then 3 cm, and so on. If the first circle has radius 3/4", the next would have radius 1 1/2", the next 2 1/4", and so on.

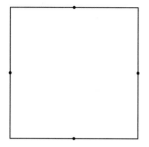 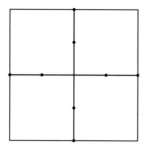 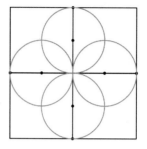

6. d. Draw first a square and mark the midpoints of its four sides (by measuring).

Draw the lines from midpoint to midpoint as you see in the image. The big square is now divided into quarters. Consider the smaller squares. Locate the midpoints of their sides by measuring (look at the image).

Draw the circles using those midpoints as center points for the circles.

Quadrilaterals, p. 20

1. You can draw several. You can draw a square with 2-in sides, or you can draw various rectangles where the one side is 2 inches and the other side is longer or shorter.

2. A quadrilateral with congruent sides and equal angles is called a *square*.

3. a. parallelogram
   c. trapezoid
   e. trapezoid
   g. kite
   i. rhombus
   b. kite
   d. rhombus
   f. parallelogram
   h. trapezoid
   j. parallelogram

4. Answers will vary. For example:

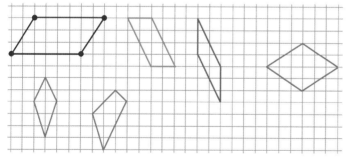

5. **Quadrilaterals Tree Diagram**

49

Quadrilaterals, cont.

6. a. Yes. A rhombus also has two pairs of congruent sides, like a kite. The two pairs just happen to be congruent.
   b. Yes. A square has two pairs of congruent sides, like a kite. Same as the rhombus, the two pairs are congruent.
   c. No. A rectangle's adjacent sides are of different lengths—unless the rectangle is also a square.
   d. Yes. A square has a pair of parallel sides (actually two of them), so it qualifies as a trapezoid, too.
   e. No. Adjacent sides of a parallelogram are of *different* lengths, so it doesn't qualify as a kite.

7. Yes, you can, by varying how tall the trapezoid is or by varying the angles. For example (the figures are not to scale):

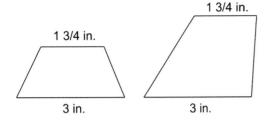

8. The adjacent sides of a parallelogram are not congruent, so it cannot be a kite or a rhombus. However, it is also a trapezoid. (From the tree diagram in answer (5), you can see that *all* parallelograms are trapezoids).

9. It is neither a kite, nor a parallelogram, nor a rhombus.

10. She can draw several different kinds: not only a square, but also rhombi of various shapes that are not squares.

11. A square, a rectangle, a parallelogram, and a rhombus all have two pairs of congruent sides and two pairs of parallel sides.

12. Quadrilaterals puzzle: YOU GOT THE SHAPES SORTED!

Equilateral, Isosceles, and Scalene Triangles, p. 24

1. Equilateral: f, g.
   Isosceles: a, c, d (and f and g)
   Scalene: b, e, h.

2.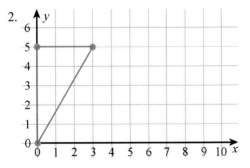

   It is a right triangle.

3.

Triangle	Classification by the angles	Classification by the sides
a.	acute	equilateral
b.	right	scalene
c.	obtuse	isosceles
d.	acute	scalene

4.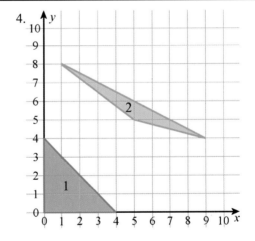

Triangle 1: right and isosceles
Triangle 2: obtuse and scalene

Equilateral, Isosceles, and Scalene Triangles, cont.

5.

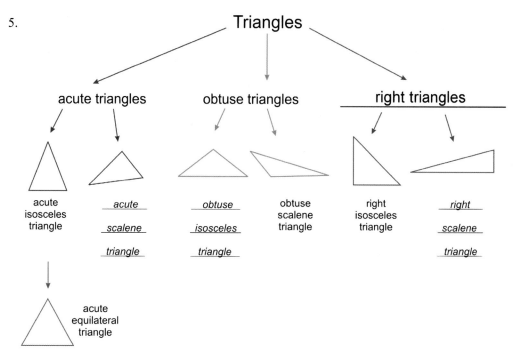

Triangles

acute triangles → acute isosceles triangle, <u>acute</u> <u>scalene</u> <u>triangle</u>

obtuse triangles → <u>obtuse</u> <u>isosceles</u> <u>triangle</u>, obtuse scalene triangle

right triangles → right isosceles triangle, <u>right</u> <u>scalene</u> <u>triangle</u>

acute isosceles triangle → acute equilateral triangle

6. a. Answers will vary. Check the student's answer. For example:

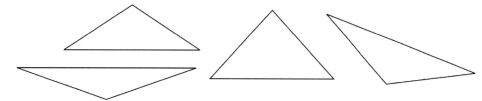

b. Answers will vary. Check the student's answer. For example:

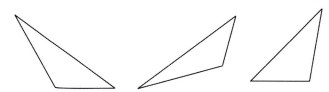

c. Answers will vary. Check the student's answer. For example:

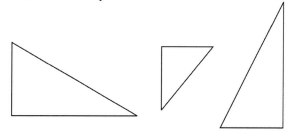

7. Answers will vary. Check the student's answer. For example:

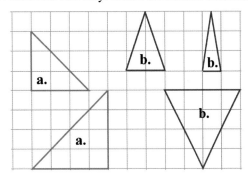

8. The angles measure 60°.

9. a. No, it cannot. The angles in an equilateral triangle measure 60°, so none of them can be a right angle.

   b. Yes, it can. For example:

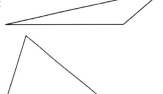

   c. Yes, it can. For example:

10. a. No, it is not possible to draw an obtuse equilateral triangle. All angles in an equilateral triangle measure 60°; therefore they are not obtuse.

    b. No, it is not possible to draw a right equilateral triangle. All angles in an equilateral triangle measure 60°; therefore they are not right angles.

    c. Yes, it is possible.

11. a. 126°, 27°, 27°.
    b. 80°, 50°, 50°.

12. a. Answers will vary. Check the student's answer. The student should have drawn two lines of the same length from the same starting point with any angle between them. The third line "closes" the triangle.
    b. The angle measurements will vary. Check the students' work. Two of the angles should measure the same, and the sum should be about 180°.

13. Students should notice that in an isosceles triangle, two of the angles have the same measure.

14. The actual size of student's triangle may vary, but it should have the same shape as this one: →

    The top angle should measure 30°.

15. a. See the image on the right (not to scale).

    b. The third side is about 7.1 cm.

    c. The top angle is the right angle. The two base angles measure 45°.

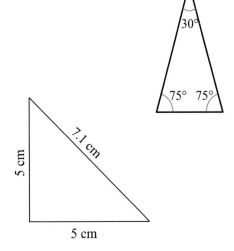

Equilateral, Isosceles, and Scalene Triangles, cont.

16. Answers will vary. There are an innumerable number of different triangles you can draw with this information, and they are not congruent. You do not even have to have the 7-cm and 3-cm sides form the obtuse angle, though it was explained that way in the hint.

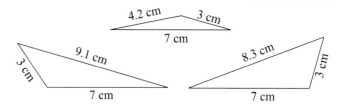

Puzzle corner.
a. The actual size of student's triangle may vary, but it should have the same shape as this one: →

b. The base angles are 65°.

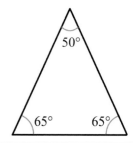

Area and Perimeter Problems, p. 29

1. a. 19 ½ sq. in. The area of the larger rectangle is 10 in. × 15 in. = 150 sq. in.
   The sides of the white rectangle are 15 − 9 ½ − 2 ¼ = 3 ¼ in. and 10 − 2 − 2 = 6 in.
   The area of the white rectangle is thus (3 ¼ in.) × 6 in = 19 ½ sq. in.
   b. The shaded area is 150 sq. in. − 19 ½ sq. in. = 130 ½ sq. in.

2. 324 cm². The total area of picture and frame is 38 cm × 29.5 cm = 1,121 cm². Subtract the width of the frame (twice, once for each side) from these dimensions to find the width and length of the picture inside the frame:
   38 cm − 2.6 cm − 2.6 cm = 32.8 cm, and 29.5 cm − 2.6 cm − 2.6 cm = 24.3 cm.
   The area of the picture inside the frame is thus 32.8 cm × 24.3 cm = 797.04 cm² ≈ 797 cm².
   The area of the frame is the difference between the total area and the area of the picture: 1,121 cm² − 797 cm² = 324 cm².

3. a. Area: Divide the figure into five rectangles. The middle rectangle has an area of 6 ft × 6 ft = 36 sq. ft. Each of the outer rectangles has an area of 6 ft × 4 ft = 24 sq. ft. In total, the area is then 4 × 24 sq. ft + 36 sq. ft = 132 sq. ft.
   b. Perimeter: 4 × 6 ft + 8 × 4 ft = 24 ft + 32 ft = 56 ft.

4. 10 cm. Subtract the long side twice from the perimeter: 42 cm − 11 cm − 11 cm = 20 cm. This is now twice the length of the short side. Take half of that to find the length of the short side, 10 cm.

5. The areas of the four component rectangles are: 6 × 12 = 72 cm²; 9 × 6 = 54 cm²; 6 × 12 = 72 cm²; and 6 × 9 = 54 cm².
   So, the total area is 72 cm² + 54 cm² + 72 cm² + 54 cm² = 252 cm².
   The perimeter is 3 + 3 + 6 + 6 + 9 + 6 + 6 + 9 + 3 + 6 + 6 + 9 + 6 + 6 = 90 cm.

6. 125 ft. Again, subtract the given side *twice* from the perimeter, and you will have twice the other side:
   910 ft − 330 ft − 330 ft = 250 ft. Half of 250 ft is 125 ft.

7. a. (1/2 mi) ÷ 4 = 1/8 mi.
   b. To convert from miles to feet, multiply by 5,280: (1/8) × 5,280 = 5,280 / 8 = 660 ft.
      You can also solve this by converting the perimeter of 1/2 mile to feet, and dividing that by 4.

8. The areas of the rectangles 1, 2, 3, and 4 illustrated at right are:
   Rectangle 1: (1 + 1) × (1 + ½) = 2 in × 1 ½ in = 3 sq. in.
   Rectangle 2: (¾ + 1 + 1) × 2 = 2 ¾ in × 2 in = 5 ½ sq. in.
   Rectangle 3: (1 + ½) × ½ = 1 ½ in × ½ in = ¾ sq. in.
   Rectangle 4: 1 × (½ + 2) = 1 in × 2 ½ in = 2 ½ sq. in.

   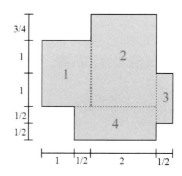

   The total area is the sum:
   3 sq. in. + 5 ½ sq. in. + ¾ sq. in. + 2 ½ sq. in. = 11 ¾ sq. in.

   The perimeter is:
   1 ½ + ¾ + 2 + 1 ¾ + ½ + 1 ½ + ½ + ½ + 2 ½ + 1 + 1 + 2 = 15 ½ in.

Volume, p. 32

1. a. V = 6 cubic units
   b. V = 16 cubic units
   c. V = 12 cubic units
   d. V = 32 cubic units
   e. V = 36 cubic units
   f. V = 28 cubic units
   g. V = 12 cubic units
   h. V = 20 cubic units
   i. V = 16 cubic units

2. a. V = 3 in³ or 3 cu. in.
   b. V = 30 ft³ or 30 cu. ft.
   c. V = 12 cm³
   d. V = 6 m³
   e. V = 13 cm³
   f. 19 in³ or cu. in.

3.

	a.	b.	c.	d.
Cubes in the bottom layer	8	12	21	6
Height	4	4	3	6
Volume	32	48	63	36

4. There are 10 × 3 × 3 = 90 cubes that fit into the box. So, the volume of the box is 90 cubic inches.

5.

	a.	b.	c.	d.
Width:	3	3	3	4
Depth:	3	2	2	4
Height:	2	3	6	4
Volume:	18	18	36	64

6. a. V = 3 × 3 × 1 = 9 cubic units
   b. V = 3 × 2 × 2 = 12 cubic units
   c. V = 3 × 2 × 5 = 30 cubic units
   d. V = 4 × 4 × 3 = 48 cubic units

7. Check students' answers. The orientation of the prisms may vary. For example:

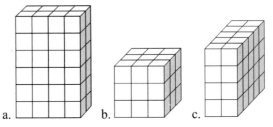

   a.   b.   c.

8. Both are correct. The two prisms just have a different orientation. Both have the same volume, and both have dimensions of 5, 3, and 2 units.

9. c. 2 × 2 × 2 + 2 × 2 × 5

Volume of Rectangular Prisms (Cuboids), p. 37

1. The order of factors in the multiplications may vary.

   a. V = _5 in._ × 7 in. × 2 in. = 70 in³

   b. A = 25 ft × 20 ft = 500 ft²

   V = 25 ft × 20 ft × 9 ft = 4,500 ft³

2. a. 20 cm × 30 cm × 50 cm = 30,000 cm³
   b. 16 in² × 6 in. = 96 in³

3. Its depth is 2 cm, because 5 cm × 3 cm × 2 cm = 30 cm³.

4. Check students' answers. The measuring could be done in inches or in centimeters, and the volume will be correspondingly in cubic inches or cubic centimeters.

5. Total volume is 28,000 cm³. The bottom box:
   V = 50 cm × 20 cm × 20 cm = 20,000 cm³.
   Top box: V = 20 cm × 20 cm × 20 cm = 8,000 cm³.
   Notice the depth of the top box is the same as the depth of the bottom box (20 cm). Total volume is found by adding and is 28,000 cm³.

6. a.

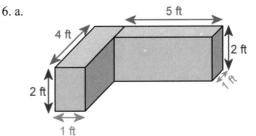

   b. The volume is 1 ft × 4 ft × 2 ft + 1 ft × 5 ft × 2 ft = 8 ft³ + 10 ft³ = 18 ft³.

7. a. Answers will vary. Check the students' work. One possibility is: Width 4 in., height 4 in., and depth 4 in. Another is: Width 2 in., height 8 in., and depth 4 in.

   b. Answers will vary. Check the students' work. One possibility is: Width 12 cm, height 10 cm, and depth 10 cm. Another is: Width 50 cm, height 8 cm, and depth 3 cm.

8. Its height is 6 inches, because 5 in. × 6 in. × 3 in. = 180 in³. Or, you can use division and solve 180 ÷ (5 × 6) = 6.

Volume of Rectangular Prisms (Cuboids), cont.

9. The total volume is 936 m³. The volume of the 'bottom' part is 12 m × 8 m × 6 m = 576 m³. The volume of the upper part is 5 m × 8 m × 9 m = 360 m³. The total volume is 576 m³ + 360 m³ = 936 m³.

10. The volume of the lowest part:
12 cm × 16 cm × 8 cm = 1,536 cm³.

The volume of the tallest part:
10 cm × 20 cm × 8 cm = 1,600 cm³.

The volume of the third part:
30 cm × 10 cm × 8 cm = 2,400 cm³.

The total volume is:
1,536 cm³ + 1,600 cm³ + 2,400 cm³ = 5,536 cm³.

11. a. 40 cm × 27 cm × 22 cm = 23,760 cm³.
    b. Since it is 1/4 full, the volume of water is 23,760 cm³ ÷ 4 = 5,940 cm³.

12. a. 12 ft × 18 ft × 8 ft = 1,728 cu. ft.
    b. The room loses 12 ft × 18 ft × 1 ft = 216 cu. ft. of volume.

13. a. The total volume is 107,400 in³.
    Part on the left:
    V = 38 in. × 30 in. × 50 in. = 57,000 in³.
    Part on the right:
    V = 40 in. × 30 in. × 42 in. = 50,400 in³.
    The total volume is 107,400 in³.

    b. 62.153 ft³.

Puzzle corner. The edge length of the larger cube is 10 in. Therefore, the edge length of the smaller cube is 5 in., and its volume is 5 in. × 5 in. × 5 in. = 125 cubic inches. The total volume of the two boxes is then 1,125 cubic inches.

A Little Bit of Problem Solving, p. 41

1. a. The area is 484 m² (22 m × 22 m).
   b. Square number 50, which has a side 100 m long.

Square number	Side Length	Area
1	2 m	4 m²
2	4 m	16 m²
3	6 m	36 m²
4	8 m	64 m²
5	10 m	100 m²
6	12 m	144 m²

2. a. Student sketches may vary, but should show a small rectangle inside a large rectangle, and the dimensions marked.

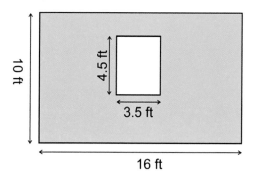

b. The area of the window is 3.5 ft × 4.5 ft = 15.75 ft².

c. The area of the actual wall is
16 ft × 10 ft − 15.75 ft² = 160 ft² − 15.75 ft²
= 144.25 ft².

2. d. Since 1 gallon covers 350 sq. ft., it is obviously enough to cover the whole wall.

e. 1 quart will cover 1/4 of 350 ft², which is 87.5 ft². Then, 2 quarts will cover double that, or 175 ft². So 2 quarts is enough.

3. a. V = 4 ft × 3 ft × 3 ft = 36 cu. ft.
   b. 32 cu. ft.

4. 2 inches

5. 40/105 = 8/21

6. (5 ft × 4 ft)/(15 ft × 20 ft) =
   (20 sq. ft)/300 (sq. ft) = 1/15.

7. a. Check students' triangles. The triangle should look like this (image not to scale):

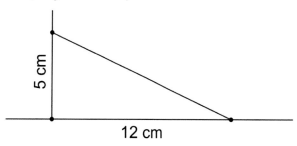

b. 30 cm. (The third side of the triangle should measure 13 cm, or close.)

Review, p. 43

1. a. Angles 64°, 64°, and 52°. It is an acute isosceles triangle.
   b. Angles 30°, 125°, and 25°. It is an obtuse scalene triangle.

2. a. Image on the right (not to scale):
   b. The top angle is 80°.
   c. Perimeter: 70 mm + 54 mm + 54 mm = 178 mm

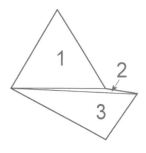

50° 50°

7 cm

3. Starting from the top left corner and going clockwise, the perimeter is:
   (¾ + ½ + 2) + (1 + 1 + ½ + ½ + ½ + ½) + (2 + ½ + ½ + ½ + ¾) + 1 + 1

   = 3 ¼ + 4 + 4 ¼ + 2 = 13 ½ inches.

   The area is: (2 in. × ¾ in.) + (3 in. × 2 ½ in) + (½ in. × ½ in.) = 1 ½ in² + 7 ½ in² + ¼ in² = 9 ¼ in².

4. a. rhombus   b. rectangle   c. square   d. kite   e. parallelogram   f. scalene quadrilateral   g. trapezoid

5. a. rectangle   b. rhombus   c. trapezoid

6. a. a pentagon   b. c. d. There are several ways to draw the diagonals:

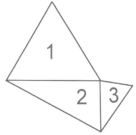
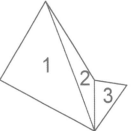

Triangle 1: Acute isosceles triangle
Triangle 2: Right scalene triangle
Triangle 3: Acute scalene triangle

Triangle 1: Acute scalene triangle
Triangle 2: Obtuse scalene triangle
Triangle 3: Acute scalene triangle

Triangle 1: Acute isosceles triangle
Triangle 2: Obtuse scalene triangle
Triangle 3: Obtuse scalene triangle

7. Answers will vary. Check students' answers. For example: a.   b.

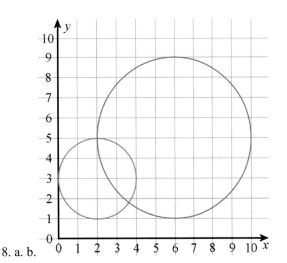

8. a. b.

9. a. The figure has 4 × 5 × 2 = 40 cubes, so its volume would be 40 cubic inches.
   b. In this case, each little cube would have a volume of 2 in. × 2 in. × 2 in. = 8 cu. in. There are 40 cubes, so their total volume is 40 × 8 cu. in. = 320 cu. in. Or, you can calculate the volume by first calculating the three dimensions: the length is 10 inches, the depth is 4 inches, and the height is 8 inches, so the volume is then 10 in. × 4 in. × 8 in. = 320 cu. in.

10. 4 cm. The area of the bottom is 2 cm × 4 cm = 8 cm^2, so the missing dimension is 32 cm^3 ÷ 8 cm^2 = 4 cm.

11. Three boxes. One box has a volume of 6 in. × 3 in. × 2 in. = 36 cubic inches. You will need three of them to have a total volume of 108 cubic inches.

Puzzle corner. The edge length of the cube must be 4 cm. Therefore, its volume is 4 cm × 4 cm × 4 cm = 64 cm^3.

Appendix: Common Core Alignment

The table below lists each lesson, and next to it the relevant Common Core Standard.

Lesson	page number	Standards
Review: Angles	11	4.MD.5
Review: Drawing Polygons	13	4.G.1
Star Polygons	15	
Circles	17	5.G.1
Quadrilaterals	20	5.G.3 5.G.4
Equilateral, Isosceles, and Scalene Triangles	24	5.G.4
Area and Perimeter Problems	29	4.MD.3
Volume	32	5.MD.3 5.MD.4
Volume of Rectangular Prisms (Cuboids)	37	5.MD.4 5.MD.5
A Little Bit of Problem Solving	41	5.NBT.7 5.OA.3 5.MD.5
Review	43	5.MD.4 5.MD.5 5.G.1 5.G.4

Made in the USA
Monee, IL
16 March 2020